IF ROCKS COULD TALK

B. Jane Bush

DALE SEYMOUR PUBLICATIONS

Product Manager: Bev Dana

Product Editor: Lisa Yount

Production/Manufacturing Manager: Janet Yearian

Production/Manufacturing Coordinator: Leanne Collins

Design Manager: Jeff Kelly

Text and Cover Design: Diane Goldsmith, Square Moon Productions

Illustrations: B. Jane Bush

This book is published by Dale Seymour Publications, an imprint of the Alternative Publishing Group of Addison-Wesley.

Order number DS31200

ISBN 0-86651-617-4

1 2 3 4 5 6-MA-97 96 95 94 93 92

DALE
SEYMOUR
PUBLICATIONS
P.O. BOX 10888
PALO ALTO, CA 94303

NOTE FROM THE AUTHOR

The research for **If Rocks Could Talk** took about three years, and my whole family became involved in it. My husband, Vern, and my three daughters, Sarah, Rachel, and Andrea, all helped to collect data.

We visited prehistoric Native American Indian rock art sites in Arizona, Colorado, Idaho, Nevada, Utah, and Wyoming. Vern and Andrea took more than 3,000 photographs while I completed about 600 drawings. I made the drawings as accurately as possible, leaving out graffiti, bullet holes, or other damage that unfortunately has been done to some of the art.

URARA (Utah Rock Art Research Association) members were also very helpful in sharing information that helped with the research for this project.

Jane Bush

CONTENTS

IF ROCKS COULD TALK

The rock surfaces of the deserts and canyons of the Southwest are filled with prehistoric drawings made by American Indians. These images tell us something about people who lived perhaps thousands of years ago. So far, much of what the drawings say remains a mystery, but experts are learning to read a few of the pictures. As we begin to understand the meanings of these images, we see that perhaps rocks can talk.

THE FIRST AMERICANS

There is some evidence to indicate people lived in the Americas as early as 25,000 years ago, although the experts do not agree on the earliest dates. Many anthropologists (scientists who study the development and culture of human beings) believe that people who crossed a land bridge from Asia 12,000 to 15,000 years ago were the predecessors of the American Indians. There is also some evidence of Asian migrations by sea about 3,000 years ago.

Anthropologists, archaeologists (scientists who study objects made by early peoples), and other scientists have been able to tell us about these ancient people. Scientists have found evidence to suggest that the earliest people were hunters and gatherers. The people wandered from place to place, living on small animals they caught and on plants they collected. The climate was not as dry then as it is now, and many plants were available to these early people that later Native Americans did not have. The early people are sometimes called the "Basket makers" because they produced large baskets to hold seeds and plants that they had gathered.

After this hunter-gatherer period, people settled into relatively permanent dwellings. They began to plant corn, beans, and squash. They kept animals such as turkeys, wove cloth and baskets, and made pottery.

HISTORIC AND PREHISTORIC INDIAN TRIBES OF THE SOUTHWEST

WA

Blackfoot

Yakima

Flathead

MT

Skokomish

Nez Perce

Crow

OR

ID

Klamath

Bannock

Shasta

Shoshone

Paiute

Great Basin

Fremont

WY

Cheyenne

Washo

Gosiute

Arapaho

NV

Ute

CA

UT

CO

Paiute

Anasazi **Four Corners**

Navajo

Apache

Zuni

Pueblos

Hopi

Mohave

AZ

NM

Hohokam

Mogollon

WHO MADE
SOUTHWESTERN ROCK ART?

Early Indians may have started making rock art fairly soon after they came to North America. Experts think the oldest Southwestern rock art may be 10,000 years old. Rock art has been found in 41 of the 50 states in the United States. All the examples of rock art in this book come from the Western United States.

Many examples of early American rock art come from an area called the Great Basin. This large area covers parts of what are now Utah, Nevada, Idaho, Oregon, and California. It is called the Great Basin because it is shaped roughly like a large bowl. All of its rivers and streams flow into Utah's Great Salt Lake. People have lived in the Great Basin continuously for 12,000 years.

Scientists think two groups of American Indians called the Fremont and the Anasazi made much of the rock art in the Southwest. The Fremont Indians lived between A.D. 300 and A.D. 1350 in what are now Utah and Colorado. Settling in small villages of 20 to 30 people, the Fremont depended on farming as well as hunting and gathering. They planted corn, squash, and beans, and they raised turkeys and perhaps goats.

The Anasazi lived in the Southwest at the same time as the Fremont Indians. They stayed in an area now called Four Corners, where the borders of four states (Arizona, Utah, Colorado, and New Mexico) meet.

PREHISTORIC GROUPS OF THE SOUTHWEST

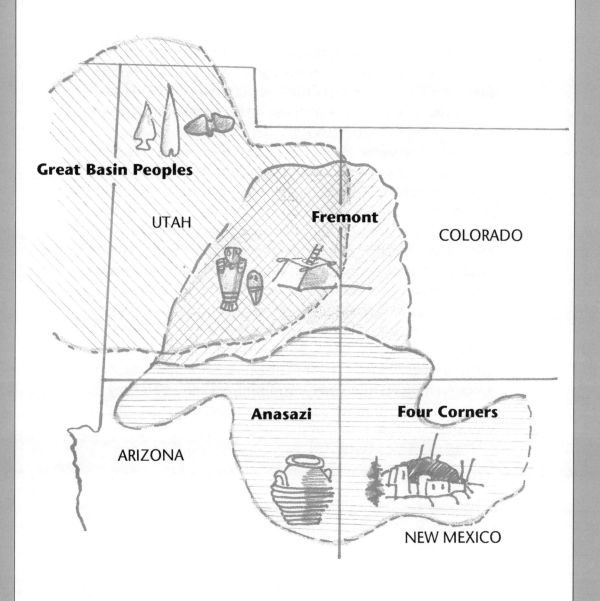

Great Basin Peoples

UTAH

Fremont

COLORADO

Anasazi

Four Corners

ARIZONA

NEW MEXICO

HOW DID ANCIENT SOUTHWESTERN INDIANS LIVE?

Much of our information about the lives of ancient peoples such as the Fremont and Anasazi comes from ruins of their villages and graves and from artifacts found there. Artifacts are objects such as pottery or tools made by human beings. Some of these artifacts have been taken to museums and laboratories, where archaeologists examine and preserve them.

To provide clues about when ancient Indians lived, archaeologists date artifacts. The most common methods of dating are carbon dating, dendrochronology, and stratigraphy. Carbon dating measures the amount of radioactive carbon left in artifacts made from things that once were alive, such as cloth (made from plants) or bone tools. This process can date objects because radioactive carbon breaks down at a steady rate. Dendrochronology involves studying yearly growth rings in pieces of wood, such as house beams. By comparing the rings to rings in ancient trees whose ages are known, scientists can date wooden artifacts accurately. Stratigraphy allows archaeologists to date objects by studying layers of debris. The oldest objects are found in the bottom layers.

These pieces of pottery were found in a Fremont village in Sevier County, Utah. Archaeologists have dated them between A.D. 500 and A.D. 1350.

The most common decorations found on Fremont pottery are spirals or stair-step designs.

Other artifacts found in ancient Southwestern villages include arrowheads and spear points, needles made from bone, and grinding tools called **manos** and **metates**. Archaeologists also have found many stone balls, ranging in size from marbles to golf balls. The scientists think these balls may have been used to grind small amounts of materials such as pigments (dry coloring substances) for paint. Some balls may have been toys or gaming pieces. Several clay figurines also have been found, ranging from three to seven inches high.

Other artifacts:

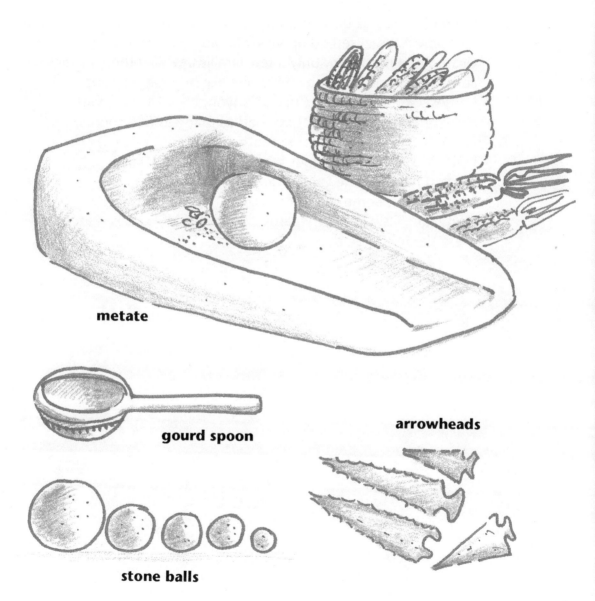

metate

gourd spoon

arrowheads

stone balls

Archaeologists have also learned about Southwestern Indians' homes. Pit houses were the most common dwellings of the northern Fremont Indians. The houses were partly buried, with only the roofs above ground. The roofs were made of branches covered with soil. An opening in the roof served as both entrance and smoke hole.

Some Anasazi settled on valley floors. Many of the villages had homes for only a few families, while others contained stone buildings that were like huge apartment houses. One building in Chaco Canyon, New Mexico, was five stories high and may have had as many as 600 rooms.

Other Anasazi (and some southern Fremont as well) built their homes on ledges in the sides of cliffs. They planted fields on the valley floor or on the flat cliff tops, called mesas. Some of the Anasazi cliff dwellings housed many families. One of the largest of these cliff houses is at Mesa Verde, Colorado.

Fremont pit house

Pueblo Bonito, Chaco Canyon

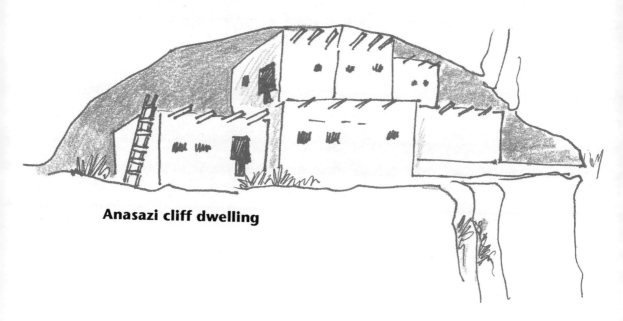

Anasazi cliff dwelling

WHAT KINDS OF ROCK ART HAVE BEEN FOUND?

Most of the rock art of the Southwest can be divided into two groups, depending on how it was made. Painted images are called **pictographs**. Images pecked or scratched into rock surfaces are called **petroglyphs**. The examples of rock art in this book are all pictographs or petroglyphs.

Other rock art forms appear in some parts of the United States and South America. Large mounds or hills made in animal forms are called **geoglyphs.** In some areas, large images were scraped into the soil. All the small rocks and plants were then removed from the images. These images are called **intaglios**.

Pictograph

Petroglyph

WHEN AND HOW WAS ROCK ART MADE?

Archaeologists have just recently begun to study Southwestern rock art. Much of the information we have about this art comes from amateur scientists who became interested in understanding and preserving the art. These people photographed, drew, and carefully documented thousands of panels of rock art throughout the Southwest.

Rock art is difficult to date because carbon dating, dendrochronology, and stratigraphy cannot be applied to rocks. Style, techniques, and subject matter are the most accurate ways to date rock art.

The rocks themselves give some clues to the age of art that appears on them. As water seeps through rocks, it often leaves behind a dark, shiny mineral coating called "desert varnish." If desert varnish has partly covered an image, the image is likely to be relatively old. Later images often overlap earlier ones. Artists who made one picture on top of another may have been trying to capture the power of the older image and add it to the newer one.

The figure on the petroglyph on the opposite page is called "Grandmother Spider" or "Spider Woman." Grandmother Spider is mentioned in the oral history of many Indian tribes. Most of these tribes think she helped to weave the world or helped the people find their way to this world.

Some scientists think that this petroglyph, which was found in Wyoming, was made by the early Shoshone Indians. These experts date the petroglyph at A.D. 1300-A.D. 1350 because carbon dating shows that artifacts found in a nearby Shoshone village were made at about that time. However, there is no way to say for certain that the petroglyph came from the same period as the artifacts. It may have already been there when the Shoshone people built their village, or it may have been carved at a later date.

Because rock art cannot be dated reliably by other means, archaeologists usually date rock art by its style. The timeline on pages 16 and 17 shows the major styles of Southwestern rock art and the approximate dates that archaeologists assign to each style.

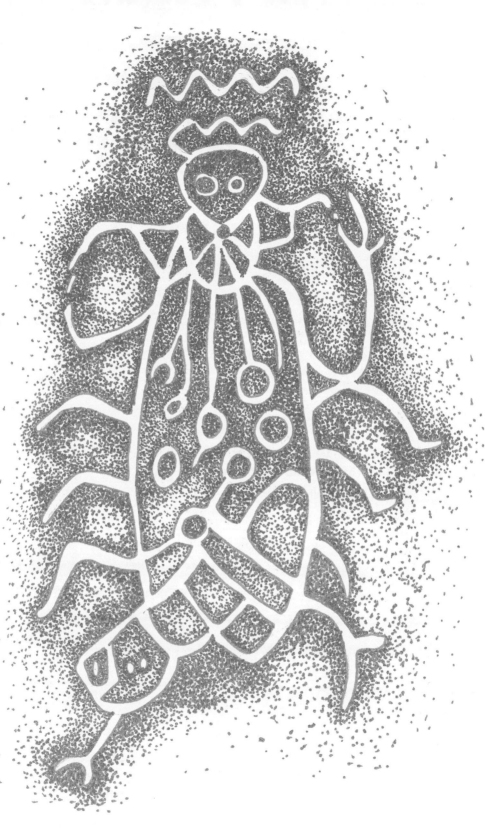

MAJOR ROCK ART STYLES
OF THE SOUTHWEST

DESERT ARCHAIC
(Simple abstract designs)
6000 – 3000 B.C. (?)

GLEN CANYON
(Line drawings)
3000 – 0 B.C. (?)

BARRIER CANYON
(Painted figures)
500 B.C. – A.D. 500

FREMONT AND
NORTHERN ANASAZI
(Stylized humanlike figures)
A.D. 300 – A.D. 1350

FREMONT, ANASAZI,
PUEBLO AND OTHERS
(Representational/realistic)
A.D. 900 – A.D. 1900

The oldest style of Southwestern rock art is called Desert Archaic. **Archaic** means "very old." Desert Archaic art, some of which may be 10,000 years old, has been found in the Great Basin.

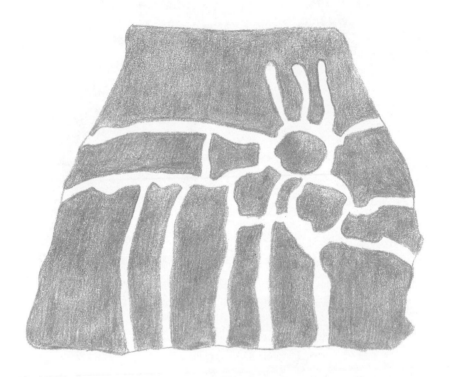

Some rock art of the Great Basin is found on large boulders sitting in the middle of a flood plain. This petroglyph from southwestern Idaho is an example.

Southwestern Indians pecked the earliest petroglyphs with fairly round rocks. Round rocks made shattered dents in large rock surfaces. Most of these early images are simple, abstract designs. They do not appear to represent people, animals, or objects.

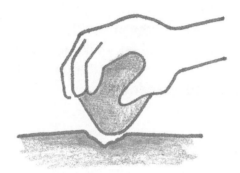

Later people fashioned pointed chisels, which were used with hammerstones. The sharp points of the chisels made deeper, clearer dents than the round rocks did. These new tools gave the artists more control and let them create images with more detail.

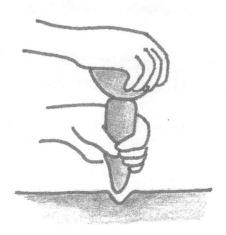

Southwestern Indians began to make painted images or pictographs as early as 500 B.C., or perhaps even earlier. Artists used sticks or brushes made from leaves of the yucca plant to paint some pictures. For others, they smeared the paint with their fingers or blew it through hollow reeds. Pigments for the paints came from crushed colored soils or from vegetable dyes. Rust red was the most common color, followed by white, black, and ocher (brownish yellow). Animal fat, milk, or blood was often added to the pigments to make the paints easier to spread and more permanent.

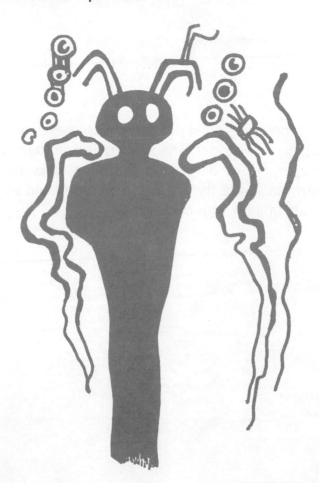

The rock art with the most detail comes from later periods, when the artists had better tools. Indians in the Southwest were trading with the Spanish by the early 1600s, and through such trade the Indians gained metal tools that could be used in making rock art. In the late 1800s a Ute artist used a railroad spike to carve this petroglyph of an owl. Many experts believe the Utes are descended from the Fremont.

Indians in some parts of the Southwest still make rock art images. The Zuni, Hopi, Ute and perhaps other tribes continue to use this age-old art form.

This is a modern Zuni pictograph showing "Na'le," the deer mask.

Today some Indian artists continue to use designs from old petroglyphs in their pottery or other art. "Kokopelli," the flute player, is one of these designs. Here Kokopelli appears on a petroglyph and is painted on a modern Acoma pot.

WHAT IMAGES APPEAR IN ROCK ART?

Animals, important in the lives of early Native Americans, are the most common rock art images. Early Indians depended on animals for food and as sources of raw materials for clothing, shelter, and tools. Perhaps you can think of more ways animals could have been used.

This chart shows some of the uses that early Native Americans found for different parts of animals.

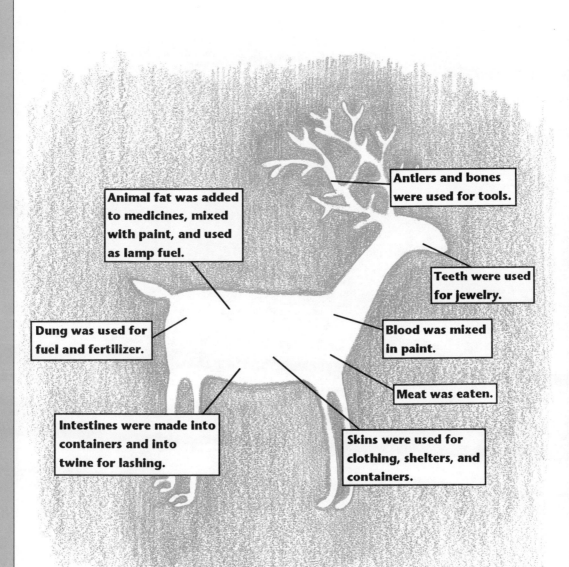

Antlers and bones were used for tools.

Animal fat was added to medicines, mixed with paint, and used as lamp fuel.

Teeth were used for jewelry.

Dung was used for fuel and fertilizer.

Blood was mixed in paint.

Meat was eaten.

Intestines were made into containers and into twine for lashing.

Skins were used for clothing, shelters, and containers.

Many rock art animals are easily recognized. Bighorn sheep and goats are the animals most commonly found in rock art. Deer, bison, antelope, turkeys, cranes, snakes, and centipedes are among the other animals portrayed.

To invite rain, people often carved frogs, water snakes, salamanders, insects called water skippers or water striders, and other water animals on rocks near their villages.

The Hopis call this horned water snake "Kolowasi." In this petroglyph it is curled into a spiral. Spirals and concentric circles are often thought to be water symbols, perhaps representing the ripples on the surface of a pond.

Many birds appear in rock art.

The Spanish brought horses to North America about 400 years ago. This Ute petroglyph containing a horse is only 100 to 200 years old. It is found in Nine Mile Canyon, Utah.

Rock images include many human figures. There are dancers, hunters, warriors with shields and spears, men on horseback, and women with children. This petroglyph is called the Fremont Family.

Other images are unfamiliar. Ghostlike figures appear to float on air. Some figures seem to wear helmets like astronauts. Some images are abstract, with meanings that remain unknown.

WHY DID PEOPLE MAKE ROCK ART?

Why did ancient Americans make rock drawings? What was the relationship between their culture and the images they used? The images may have been part of religious ceremonies or of magic evoked by the tribes' shaman-priests. The images may have been records of events or information used for the education of the tribe. They may have been symbols marking the territory belonging to a large family group or clan. Some calendars may have been made simply to pass the time while tending flocks. There is evidence to support each of these possible reasons for making rock art.

Certain rock art images appear to tell stories. Perhaps a hunter may be bragging, "I shot a six-point buck, and it took only one arrow." In another picture, hunters may be telling about meeting a bear bigger than all six men.

Other images may have conveyed simple messages rather than telling stories. Every culture uses visual images to deliver messages. The modern road sign with a deer warns motorists to watch for deer or other animals on the road. The Fremont deer, made 1,000 years earlier, is similar. It may mark a trail used by deer, or it may show a particular hunting incident.

The modern sign of the flag carrier tells drivers to look for road construction ahead and be prepared to stop. The figure in this Fremont petroglyph seems to be carrying a water jar, perhaps indicating that a village or clan was migrating.

Some glyphs are thought to be maps. They may give directions to hunting areas or ceremonial spots. Some experts think this glyph is a map showing the junction of the Green and Yampa rivers, near the Colorado-Utah border. The road map shows the rivers joining in a pattern similar to the one on the petroglyph.

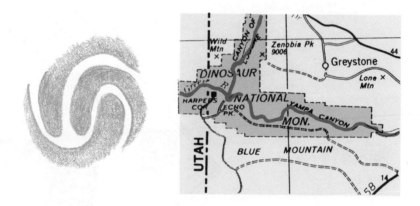

Some glyphs are "locator glyphs." They tell travelers how to find water or more important glyphs not easily seen from the path. This glyph points to another, more important panel. It tells the observer to go around the top of the panel on which this glyph is carved and look in the direction of the two pointers.

This glyph marks a natural basin where water may collect.

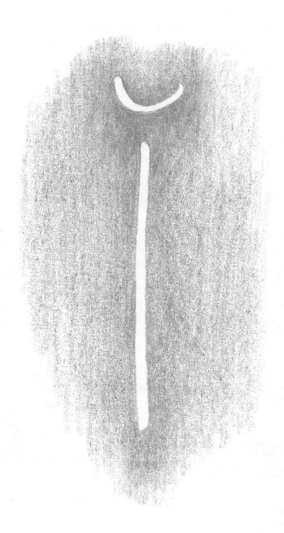

Some rock art may have been part of religious traditions. Many painted images are found in spots where rock formations surround an open space, making a natural bowl or amphitheater where people could gather for ceremonies. These pictographs, which are thought to show magical beings, are larger than life. Some drawings are eight to twelve feet tall.

Some petroglyphs showing animal images also have religious meanings. This petroglyph shows a dog with his front paws outstretched to the morning sun as if praying. The sun was very important in the religion of many Southwestern tribes. The tribespeople believed the sun to be the father and the earth to be the mother of all living things.

Many petroglyphs are calendars. Some tribes of the Southwest divide their year into 8 parts, with about 45 days in each. The dividing days in this calendar are the summer and winter solstices (the year's longest and shortest days), the spring and fall equinoxes (times when day and night are of equal length), and the midpoints between these 4 times. The midpoints are called cross-quarter days. Certain petroglyphs showed when some or all of these 8 days occurred.

Solar interactor petroglyphs were also a type of calendar. Research now in progress indicates certain petroglyphs were placed so the sunlight made shadows that completed the petroglyph designs on certain important days. These petroglyphs are called solar interactors because the sun interacts with them in this special way. The days on which the sun interacted with the petroglyphs were usually solstices, equinoxes, or cross-quarter days.

This goat figure is one example of a petroglyph calendar that interacts with the sun to mark important days. Shadows line up with the five lines through the figure's body on both solstices, both equinoxes, and the cross-quarter days. The first and fifth lines mark the solstices. The line in the middle marks both equinoxes. The second and fourth lines mark cross-quarter days. When the sun reaches a solstice mark, it begins moving through the figure again in the opposite direction. The sun therefore moves through the figure twice, once in each direction, during each year.

Other glyph calendars are highlighted by the sun on only one or two days a year, letting the people know when it was time to begin planting corn or squash.

The petroglyphs on the next page are also solar interactors. Shadows cut the top two spirals in half on the equinoxes, telling the Fremont and Anasazi that the days now would start becoming longer or shorter.

The glyph in the lower left corner shows a shaman or priest. On certain days of the year a shadow makes the glyph look as if light is coming out of the shaman's mouth. People might say that the shaman "speaks with light."

The glyph in the lower right corner of the next page is an observation locator. If an observer stands directly under the arch and looks east, the sun will appear at a prominent spot on the horizon at solstice, equinox, or a cross-quarter day. The drawing on the bottom of this page shows how an observer might use this glyph. Like other solar interactors, this observation locator acts as a calendar.

Some animal glyphs seem to represent guardian spirits. Ancient Indian tribes often encouraged young people to go on a quest to seek a guardian animal spirit. A young person would stay alone in the wilderness without eating until an animal guardian revealed itself in a vision or dream. In some cases a family or a village would choose an animal for a guardian, and this animal would be its clan sign. Many petroglyphs are probably clan signs that identify hunting areas belonging to the Bear Clan, the Turtle Clan, and so on.

These petroglyphs are thought to be clan signs for the Bear, Corn, Turtle, Lizard, Beetle, Rabbit, and Badger clans.

Shields are common images in rock art. Real Southwestern Indian shields were made from animal hides or woven like baskets. The shields in this petroglyph seem like x-ray pictures because you can see the figures through them. One of the figures on the right was made upside down to show that the man it represented had died.

Some painted images mark burial spots. A warrior's grave may be marked with a painted shield.

HOW CAN WE PROTECT ROCK ART?

Today our heritage of Southwestern rock art is in danger. Some rock art panels are being destroyed by natural weathering, erosion, and plants growing across the rock surfaces. People, however, cause the worst damage.

The first people who studied rock art often harmed the art accidentally. Some of these people put chemicals or chalk on the rocks to make the images show up better in their photographs. Others did rubbings of the glyphs, which gradually made the sandstone crumble away. These well-meaning people did not know that simply touching painted images can cause damage.

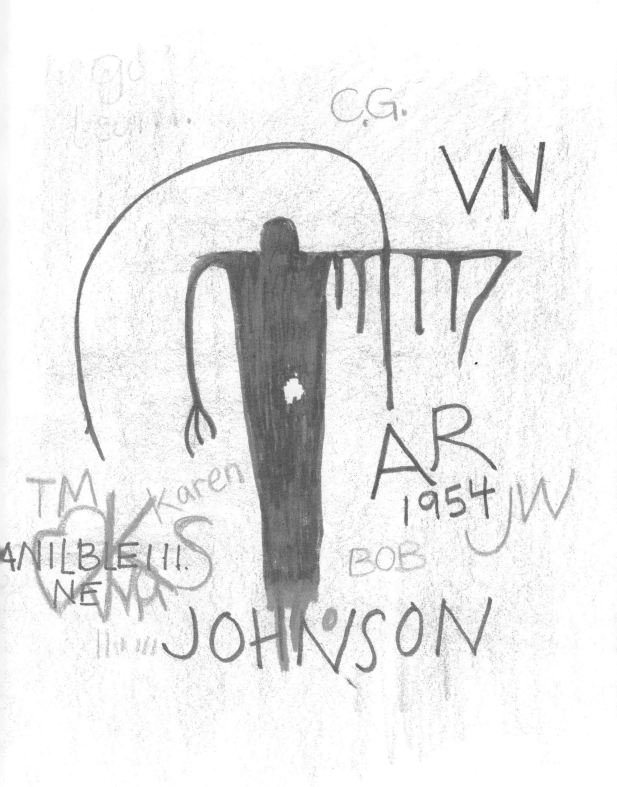

Vandals do the most harm to rock art today. They have purposely altered old images or carved their own pictures into the rocks. They have carved initials and names into even the oldest panels and used spray paint to write graffiti over the precious drawings. Some have tried to cut the rock art free so they can steal it. The damage done by vandals can never be repaired, and it has ruined some very old and beautiful rock art.

What can be done to protect rock art panels? Congress has passed several laws to protect archaeological sites and places where rock art exists. These laws make it illegal to dig for or remove artifacts or to damage any rock art. In addition, the people who do research on rock art and the National Park Service ask visitors to rock art sites to leave only footprints and take only photographs. We can all protect America's rock art heritage by following this rule.

Leave only footprints.

Take only photographs.

GLOSSARY

abstract: not easy to understand; not picturing objects that can be identified

Acoma: a group of eastern Pueblo Indians who live in New Mexico near Albuquerque

Anasazi: a group of American Indians who lived in the Four Corners area, where the borders of Arizona, Utah, Colorado, and New Mexico now meet

anthropologist: a scientist who studies human beings and the development of human culture

archaeologist: a scientist who studies objects (artifacts) made and used by ancient peoples

archaic: very old; dating from a much earlier time

artifact: an object made and used by human beings, such as a tool or a pot

badger: a burrowing animal about the size of a medium dog

carbon dating: dating an object made from formerly living materials (which contain the element carbon) by measuring the amount of radioactive carbon left in a piece of the object

centipede: a small, caterpillarlike animal with many legs

clan: a large family group that claims to be descended from a single ancestor

cross-quarter day: the day that falls exactly halfway between a solstice and an equinox

dendrochronology: a method of dating wooden objects by comparing growth rings in the wood with growth rings in trees whose age is known

desert varnish: a dark, shiny coating of iron and other minerals, deposited on rock surfaces when water seeps through them

equinox: when day and night are of equal length, which occurs once in spring and once in fall

Four Corners: an area where the borders of Arizona, Utah, Colorado, and New Mexico meet

Fremont: an American Indian tribe that lived between A.D. 300 and A.D. 1350 in what are now Utah and Colorado

geoglyph: a mound or hill made into an animal form

glyph: shortened form of petroglyph; carved or pecked image

Great Basin: region of interior drainage between the Sierra Nevada and Wasatch mountains, including most of Nevada and parts of California, Idaho, Utah, and Wyoming

Hopi: the westernmost of the Pueblo Indian tribes, living mainly in northeastern Arizona; they may be descended from the Anasazi

intaglio: an image made on the ground by scraping a design into the soil and removing small rocks and plants from the outline or inside of the design

mano: a stone held in the hand and pressed against a hard surface, used for grinding corn, other grains, or other dry material such as pigments

metate: a stone with an inward-sloping surface, against which a mano is pressed during grinding

Navajo: an American Indian tribe who live in northeastern Arizona, southeastern Utah, and northwestern New Mexico

petroglyph: an image pecked or scratched into a rock surface

pictograph: an image painted on a rock surface

pigment: a colored substance used to make paint; it could be dry or liquid

pit house: a partially buried house with a roof above ground

radioactive: having atoms that break down naturally, giving off certain rays in the process of deterioration

salamander: a small, lizardlike animal that usually lives near water

shaman: a priest or specialist in healing; such people were important religious and political leaders in many American Indian tribes

Shoshone: a large American Indian cultural group living in parts of California, Nevada, Utah, Idaho, and Wyoming

solar interactor: a glyph that conveys its meaning when the sun interacts with it, casting light or shadows on important parts of the glyph

solstices: the shortest and longest days of the year (usually on June 21 and December 21)

stratigraphy: a method of dating objects by noting the layers of debris where the objects appear; usually the deeper an object is buried, the older it is

Ute: an American Indian tribe living in western Colorado and eastern Utah; many experts believe them to be the descendants of the Fremont Indians

Zuni: an American Indian group, one of the Pueblo tribes, who live in west central New Mexico, near the Arizona border